THROUGH THE ASTRONAUT'S EYE
Stories, Secrets, and Journeys Beyond Earth - A World Without Borders

The Extraordinary View of Our Planet Through the Eyes of Those Who Left It Behind

Tommy S. Manley

Copyright ©Tommy S. Manley, 2024.

All rights reserved. No part of this publication may be reproduced, distributed, or transmitted in any form or by any means, including photocopying, recording, or other electronic or mechanical methods, without the prior written permission of the publisher, except in the case of brief quotations embodied in critical reviews and certain other noncommercial uses permitted by copyright law.

Table of Contents

Acknowledgments..4
Introduction..7
Chapter 1: Dreaming of the Stars – Beginnings and Inspirations.. 10
Chapter 2: The Path to the Stars – Training and Determination... 16
Chapter 3: The Launch – An Unforgettable Ascent 23
Chapter 4: Floating in the Void – Adjusting to Life in Zero Gravity.. 30
Chapter 5: The International Space Station – Humanity's Orbiting Lab... 38
Chapter 6: The Cupola – Earth Observed from Above 46
Chapter 7: A World Without Borders – The Unity of Humanity... 55
Chapter 8: Earth's Dynamic Beauty – The Colors and Patterns of Nature.. 64
Chapter 9: Practical Use Cases and Setup Tips...... 73
Chapter 10: Troubleshooting and FAQs...................82

Acknowledgments

This book is possible thanks to the remarkable individuals who have shared their journeys and reflections on the wonders of Earth from a vantage point few will ever know. My deepest gratitude goes to each astronaut whose stories, insights, and courage enrich these pages and reveal the extraordinary view of our planet from above.

To **Reid Wiseman**, whose words echo a transformative awe that transcends borders, and **Anne McClain**, who shared the deeply moving moment of realizing our shared humanity from the depths of space.

To **Jessica Meir**, for conveying the wonder of looking at Earth with fresh eyes, and **Alvin Drew**, whose perspective inspires a sense of unity and connectedness across continents.

To **Tracy Dyson**, who described the colors, shapes, and vastness of Earth's natural beauty, and

Michael Foreman, who provided vivid accounts of life aboard the space station and the deep sense of connection he felt.

To **Don Pettit**, for sharing the endless wonder of the cosmos, and **Mike Fossum**, for the heartfelt realizations that made the Earth below seem so fragile and precious.

To **Bill McArthur**, who reminded us of the importance of a shared global perspective, and **David Saint-Jacques**, whose words bring to life the visceral beauty of Earth as seen from orbit.

To **Drew Morgan**, for recounting the power and beauty of our planet's natural phenomena, and **Karen Nyberg**, who captured the intricate and interconnected systems of life on Earth.

To **Hazza Al Mansouri**, for offering a distinct and heartfelt perspective, and **Jack Fischer**, whose enthusiasm lights up the narrative with his reflections on our shared home.

To **Nicole Stott**, whose observations bring the fragility and beauty of Earth to vivid life, and **T.J. Creamer**, who shares the profound emotions tied to such an extraordinary journey.

Finally, to **Nick Hague**, for underscoring the vastness of our universe and the incredible smallness of our own existence on this beautiful planet.

Thank you to each of these exceptional individuals for their commitment to exploration and their willingness to share their stories. Through their eyes, may we all gain a renewed appreciation for our shared home and a deeper understanding of our place within the vastness of space.

Introduction

From the vastness of space, Earth transforms into something wholly unique—a singular, glowing sphere, afloat in an endless dark sea. Seen from orbit, the familiar landscapes, sprawling cities, and drawn borders fade, merging into a unified mosaic of blues, greens, and subtle patterns. For the few who have ventured beyond Earth's gravity, this view brings a profound shift in perspective, a sense of connection to all life and a realization of Earth's fragility. It is this experience, often called the "overview effect," that has changed the lives of astronauts, awakening within them a deep responsibility to protect our shared home.

Each astronaut's journey began with a longing to explore the cosmos, a desire that grew into years of intense training, courage, and perseverance. These stories are told by individuals who left everything familiar behind to venture into the unknown, each finding something beyond their expectations. Their experiences reveal a mixture of awe, personal reflection, and transformation. They speak of an almost indescribable beauty, of seeing our planet from a vantage point where political boundaries disappear and humanity is woven together into a delicate web. They share stories of the stars, of Earth's vibrant colors and weather patterns, and of the fragile atmosphere, visible only as a thin blue line from space.

This book is a collection of those stories and a reflection on the insights that emerge when one sees Earth from above. As you journey through these pages, you'll find yourself sharing in the astronauts' emotions, sensing the vast unity they felt, and understanding their heightened sense of responsibility toward our planet. Through their eyes, we glimpse Earth as a living, breathing entity, resilient yet vulnerable, a place where all of life is interconnected. By bringing you closer to the astronauts' experiences and this newfound awareness, this book seeks to convey that profound realization—the reminder that, in the end, we are all custodians of the only home we know.

Chapter 1: Dreaming of the Stars – Beginnings and Inspirations

For many astronauts, the journey to space began not with a well-defined plan but with a child's simple wonder. Nights spent under star-filled skies stirred questions and dreams, planting seeds that would one day lead them beyond Earth's atmosphere. Some of them, as children, lay back on the grass, captivated by the thought of touching the stars or discovering the mysteries that lay beyond the visible sky. These early moments, when the cosmos felt both inviting and unknowable, lit a spark that continued to grow.

Jessica Meir recalls announcing at the age of five that she wanted to be an astronaut, a childhood statement that might have seemed fleeting but became a lifelong ambition. She would hold onto that spark through years of study, physical conditioning, and countless hours of preparation. For others, the journey was similarly sparked by curiosity, and the idea of space became a constant, even when the path seemed improbable or distant. Alvin Drew remembers watching early space missions with awe and feeling an inner pull, a magnetic attraction toward the stars that made space seem like the ultimate adventure. It wasn't just about going somewhere new; it was about venturing into a realm where humanity was only beginning to place its footsteps.

The path, however, was anything but easy. NASA's rigorous selection process demands not just skill but also resilience. The journey to becoming an astronaut required the courage to face rejection time and again. Michael Foreman applied repeatedly, each rejection letter only fueling his resolve. He recalls the humbling experience of knowing he was one of thousands of hopefuls, all competing for a limited number of slots. Similarly, Don Pettit, undeterred by multiple rounds of rejection over the course of more than a decade, continued to chase his dream, believing that persistence would eventually bring him to the stars.

The relentless dedication to their dream often meant sacrifice. Days, months, and years were invested in specialized training and education, often with no guarantee of success. But the vision of Earth from space, the hope of experiencing a world beyond the known, kept them going. Many astronauts describe this period as one of immense personal growth, learning to push past their limits and trust in the strength of their resolve. For them, becoming an astronaut was more than achieving a title; it was an evolution shaped by patience, humility, and an unwavering belief in the value of exploration. This journey is a testament to the power of childhood dreams and the extraordinary lengths to which people will go to make them a reality.

Reid Wiseman's journey to space was one marked by determination and a commitment to pursue a vision that many might have given up on. Early on, he understood that his path to the stars would be long and demanding. Each step he took, from rigorous academic preparation to military service and eventually training as a test pilot, was a building block toward a goal he'd cherished since childhood. Yet, there was never a guarantee of reaching space, and Wiseman encountered his own set of rejections. In those moments, it would have been easy to redirect his efforts elsewhere, to set his sights on a more reachable goal, but he couldn't shake his desire to see Earth from orbit. Every setback seemed only to strengthen his resolve. When he finally made it, the experience was nothing short of life-changing, affirming that his years of dedication had been worth every challenge.

Jessica Meir's path also reveals this remarkable blend of ambition and resilience. From her early years, she carried with her the certainty that she was meant for something extraordinary. The dream of becoming an astronaut was woven into her life, shaping her choices and pushing her to endure what others might have considered insurmountable. She pursued advanced studies, delved into scientific research, and devoted herself to rigorous training, knowing that each effort was part of a much larger journey. When she finally reached space, it was not only a testament to her personal achievement but also a beacon of possibility for anyone willing to follow their own dreams with the same level of dedication.

In much the same way, Alvin Drew's story demonstrates the resilience it takes to keep a dream alive. Growing up in an era when space exploration was still in its early stages, Drew watched humanity's first steps on the moon with wide-eyed fascination. For him, the desire to explore was more than a curiosity; it was a call to action. He knew he wanted to contribute to this monumental human endeavor, even when it required years of tireless training, personal sacrifices, and steadfast belief in his vision. Drew's journey, like Wiseman's and Meir's, was one of obstacles and persistence, reminding him that exploring space wasn't just a personal goal but part of a greater story of discovery and shared humanity.

These astronauts embody a universal longing for exploration, a drive that transcends individual backgrounds, circumstances, or borders. Their stories are rooted in a belief that humanity's potential lies beyond the limits we see before us. Every rejection, every barrier they faced only underscored their commitment to breaking through. In their journeys, there is a common thread: a deep, resilient spirit that pushes beyond the comfort of certainty and steps forward into the unknown. It is this timeless spirit—the desire to reach, to discover, to understand—that fuels exploration. Through their individual stories, Wiseman, Meir, Drew, and others reveal the resilience it takes to dream beyond our boundaries and inspire us to imagine what might be waiting just beyond the stars.

Chapter 2: The Path to the Stars – Training and Determination

Becoming an astronaut is a journey of unyielding rigor, both physically and mentally, demanding not only intelligence and skill but also resilience in the face of relentless challenge. NASA's selection process is an odyssey in itself, designed to identify candidates with an unbreakable spirit. For those who dream of space, the first hurdle is acceptance, which often comes after many attempts. Thousands apply, yet only a handful are chosen, and many well-qualified candidates face rejection, sometimes repeatedly, before ever getting a chance to train.

The physical preparation is uncompromising. Trainees undergo grueling fitness tests, hours in simulators that replicate the disorienting sensations of space, and underwater exercises to mimic the weightlessness they'll experience in orbit. Every task is designed to condition them for an environment that is hostile to human life. Training also includes survival exercises, like being dropped into the wilderness to practice staying alive with minimal resources—a lesson in self-reliance and problem-solving under pressure. These exercises demand not only strength but also the ability to stay calm under extreme stress. Some astronauts recall moments of near exhaustion, yet they knew that only by pushing their limits could they prepare themselves for the unpredictability of space.

Mental preparation is equally taxing. Long hours are spent studying complex systems, memorizing intricate spacecraft controls, and learning how to troubleshoot in the most unforgiving of environments. Candidates must learn to operate under pressure, aware that a single mistake could have life-or-death consequences in space. Many astronauts recount the patience and discipline required to meet these demands, knowing that the challenges would continue to intensify with each phase of training. They are constantly assessed, their endurance and ability to remain composed under pressure monitored closely. In a field where every decision matters, resilience is just as essential as technical expertise.

For many, the journey was one of years, even decades, marked by repeated rejections. Take Don Pettit, who spent over a decade reapplying, improving his skills, and persisting despite the odds. For every setback, he simply sharpened his focus and expanded his knowledge, believing that he could earn his place in space. Similarly, Mike Fossum applied five times over thirteen years before finally receiving the call that would change his life. Each rejection was both humbling and motivating, reinforcing the lesson that reaching space is not about initial success but rather about unwavering perseverance.

This process builds more than astronauts; it builds people capable of facing the unknown with confidence, adaptability, and an acute awareness of their own limits. The demands of training refine not only their physical abilities but also their character. Those who emerge from NASA's program have proven themselves capable of withstanding adversity, embodying the resilience that space exploration requires. In the end, every aspect of their journey prepares them for the realities of space, where every day presents a challenge, and survival hinges on the strength and resilience they've built over years of dedication.

Preparing for space travel is a transformative journey, one that demands not just skill but an exceptional level of resilience and an enduring passion for exploration. Each astronaut's preparation was a test of their abilities and their capacity to adapt to the unknown. It began with mastering an immense array of technical skills, from learning to navigate complex spacecraft systems to understanding life-support protocols. They absorbed volumes of information, from orbital mechanics to emergency procedures, each layer of knowledge crucial to their survival and the success of the mission. The training was as much about cultivating sharp reflexes and precise thinking as it was about learning to face the unexpected with poise.

For many astronauts, developing resilience was as vital as developing skill. Living in an environment as hostile as space means confronting the body's limitations in ways that few experience. They prepared for everything from intense G-forces during launch to the weightlessness that can disorient even the most seasoned travelers. Physical conditioning became a way of life; they trained to maintain muscle strength and cardiovascular health to counteract the toll of zero gravity. But even beyond the physical challenges, they faced the psychological demands of long-term isolation, learning to thrive in confined spaces, and forming habits to keep both body and mind resilient.

The passion for space exploration fueled their commitment through years of exhaustive preparation. Drew Morgan, like many of his fellow astronauts, found motivation in the belief that his work held significance beyond himself. He prepared by embracing not only the technical demands but also the mindset needed to face the unpredictable with optimism. Jack Fischer, too, saw each challenge as a stepping stone, a necessary part of the journey toward fulfilling a lifelong ambition. This passion transformed grueling routines into essential rituals, as each task brought them closer to realizing their dream of seeing Earth from above.

NASA's training also encouraged a deep-seated adaptability, where the intersection of skill and resilience became crucial. Astronauts prepared for scenarios that defied their usual control, from potential equipment malfunctions to limited resources. They trained in diverse environments to learn how to adapt on the fly, to think critically in moments of pressure, and to rely on their instincts as well as their training. This adaptability fostered a profound trust in their teammates, as astronauts would often need to rely on one another's expertise to navigate unforeseen challenges. The training process created a bond of mutual respect and trust, where individual skill blended seamlessly with teamwork, enhancing their collective resilience.

At the heart of each astronaut's journey was an unwavering passion for exploration. It was this passion that kept them focused through years of training, despite the physical toll and psychological strain. Passion is what sustained their resilience, fueling their drive to prepare for the unknown. For these astronauts, the journey was about more than reaching space; it was a testament to the belief that humanity's potential lies in pushing boundaries, in exploring what lies beyond. Their journey to space was shaped by a rare combination of skill, resilience, and passion—qualities that enabled them not only to endure the demands of space travel but to thrive in its challenges, becoming stronger, wiser, and more connected to the purpose of their mission.

Chapter 3: The Launch – An Unforgettable Ascent

Launch day is an experience unlike any other—a moment seared into memory by the rush of emotions, sensory overload, and the sheer force of the rocket as it comes to life beneath the astronauts. For each of them, the buildup is a mixture of anxiety, excitement, and an acute awareness that years of preparation have led to this very instant. As they are strapped into their seats, encased in their suits, and cocooned within the narrow confines of the spacecraft, there is a stillness, a silence punctuated only by the low hum of machinery and distant, muffled voices. The weight of the journey ahead presses on them, creating a moment of introspection amidst the countdown.

The tension builds in those last ten seconds, as the rocket's systems activate, each second ticking down with electrifying intensity. The ignition is instantaneous and thunderous, sending a shockwave through their bodies as the engines roar to life. Drew Morgan describes it as a force that feels almost alive, a "living, breathing beast" ready to break free from Earth's grasp. Jack Fischer recalls the relentless vibration, the rumble of the rocket that shakes them to their core, unlike anything that can be experienced on the ground. Karen Nyberg remembers the abruptness of the initial lift, the sharp jolt as the solid rocket boosters ignite and thrust them forward, as though the universe itself is giving them a powerful nudge toward the stars.

The physical sensations are intense and all-encompassing. With the rapid acceleration, they are pressed back into their seats, their bodies heavy under the strain of G-forces as the spacecraft gathers speed. The pull is constant, pressing down on their chests, making even simple movements feel as though they are pushing through a dense, invisible weight. Fischer compares the initial jolt to being rear-ended in a car accident—an unyielding force pushing against him with an intensity that is both exhilarating and overwhelming. Yet, as the rocket ascends, the roughness smooths out, and a sense of forward momentum takes over, the thrust leveling into a smoother but powerful ascent that gradually releases its hold as they break through the atmosphere.

Through the tiny windows, Earth begins to shrink below them, and the blue of the sky darkens, deepening until it becomes an infinite black. Their anticipation shifts to awe as they realize they are leaving their planet behind. For Nyberg, this transition is one of quiet wonder, her focus narrowing to the view outside as she feels the spacecraft breach Earth's atmosphere. In those moments, she finds herself unexpectedly calm, her excitement replaced by a reverent appreciation of the enormity of the journey.

The launch is an assault on the senses, a potent mix of noise, force, and adrenaline, but it is also a moment of profound realization. As they settle into the journey to orbit, these astronauts feel the weight of their ambition lifting, replaced by the lightness of a dream fulfilled. In those first minutes, they are aware that they are no longer bound to Earth's surface. They are part of something larger, part of the exploration of a vast and beautiful unknown. The rocket's thunderous departure fades behind them, leaving only the silence of space ahead—a silence that carries with it the promise of discovery.

As Earth falls away beneath them, the reality of leaving its surface begins to sink in, bringing with it a profound sense of exhilaration and the weight of their journey's significance. In those early moments of ascent, the astronauts are flooded with the awareness that their years of training, sacrifice, and waiting have finally converged into this single point of departure. The planet they've always known quickly shrinks in the window, and with it, the ordinary life they left behind. What was once theoretical and distant has become immediate, tangible. For the first time, they are among the few to step outside the boundaries of our world, to experience life from a perspective that most can only imagine.

The contrast between the intense force of the rocket and the sudden, surreal calm of space is striking. One minute, they are pinned against their seats by the rocket's unrelenting thrust, bodies heavy and senses flooded; the next, they are suspended in weightlessness, floating, with Earth an immense, vibrant sphere below them. This is the moment when the journey transforms from a calculated mission into something profoundly personal. Karen Nyberg describes the rush of emotions that accompany this shift, a quiet awe overtaking her as she gazes at the vastness around her. She is struck by the stillness, the clarity, and the infinite horizon stretching before her, a stark contrast to the contained and familiar world she knew on Earth.

Drew Morgan captures the duality of this moment well—the sense of responsibility mingled with the boundless freedom of being in space. The exhilaration of floating for the first time, unbound by gravity, amplifies the reality that they have ventured into an environment both beautiful and unforgiving. Their lives are now tied to their training, their skills, and the life-support systems aboard the spacecraft. But there is also a kind of liberation in leaving behind the constraints of Earth's gravity, a symbolic release that mirrors their journey to a realm that is both scientifically demanding and personally transformative.

Jack Fischer reflects on the sheer thrill of seeing Earth from this vantage point, a breathtaking panorama of blues, greens, and whites unfurling beneath him. The planet appears fragile, finite, and yet resilient, its curves illuminated against the vast, dark backdrop of space. In this defining moment, the magnitude of the journey becomes clear; they are no longer spectators but active participants in humanity's exploration of the unknown. For Fischer and others, this realization comes with a surge of pride and humility—a recognition of their role in something greater than themselves.

This moment marks the beginning of a life-changing experience, one that will forever alter how they view the world, humanity, and their place within it. They are now part of a select few who have seen Earth from above, who have experienced firsthand the beauty and isolation of space. It is a defining moment, one that crystallizes the gravity of their journey and the extraordinary privilege they hold. In that quiet, weightless expanse, each astronaut begins to understand that they are not merely leaving Earth behind; they are gaining a new perspective that will stay with them long after they return.

Chapter 4: Floating in the Void – Adjusting to Life in Zero Gravity

The first sensation of weightlessness is unlike anything most astronauts have ever experienced, a jarring shift that instantly turns everything familiar upside down. The physical body, accustomed to the steady pull of gravity, suddenly becomes unmoored, and simple movements feel foreign and unpredictable. For those who dreamed of floating effortlessly in space, the initial reality is a confusing blend of awe and disorientation. Reid Wiseman describes those early days of weightlessness as "horrible yet thrilling," the wonder of seeing objects hover around him tempered by the relentless queasiness that accompanied each movement. Many liken the feeling to seasickness, a deep internal sensation of imbalance as the body struggles to adapt to the lack of gravitational grounding.

Navigating this new environment requires not only physical adjustment but a complete rewiring of mental and spatial perception. Actions that were once automatic—moving from one place to another, orienting oneself within a room—now demand a conscious effort. There is no longer an up or down, only endless directions in which to float. Drew Morgan describes this sensation as both freeing and overwhelming, an experience that brings out the inner child but requires careful relearning. Moving through the spacecraft involves a delicate coordination, using handrails, walls, or subtle pushes to glide from one place to another without sending oneself tumbling in an unintended direction. Even the simple act of stopping requires control; the slightest miscalculation can leave an astronaut spinning gently in the wrong direction.

Adjusting psychologically is as challenging as the physical adaptation. In weightlessness, the brain's sense of equilibrium is constantly being tested, a disorienting experience that some struggle to overcome. It takes days, sometimes weeks, before the body settles into a new normal, and even then, the occasional wave of nausea can resurface. The mind, too, must learn to reorient itself to the unconventional freedom of floating in three-dimensional space. Jessica Meir recalls that the experience was deeply humbling, a reminder of how closely the mind and body are tied to the familiarity of Earth's gravity. This environment, where things as basic as drinking water or sleeping take on an entirely new dimension, requires astronauts to let go of old instincts and embrace a novel way of thinking.

Despite the challenges, the wonder of weightlessness quickly begins to overshadow the initial discomfort. There is an undeniable thrill in seeing an apple, pencil, or even droplets of water float serenely nearby, suspended in midair. This freedom is both fun and empowering, allowing each astronaut to move with a lightness that feels dreamlike. As the days go by, the body adapts, finding its rhythm in this unusual state, while the mind learns to trust this newfound balance.

Gradually, floating becomes second nature, a tool that allows them to explore the spacecraft with a fluidity impossible on Earth.

For each astronaut, adjusting to weightlessness becomes a part of their personal journey. It demands patience, resilience, and a willingness to redefine the boundaries of their physical and mental comfort zones. And as they settle into life in this gravity-free environment, the sense of disorientation fades, replaced by a quiet confidence and the sheer joy of floating freely in the vast expanse of space.

Life aboard the International Space Station (ISS) is a blend of routine tasks and extraordinary adaptations, a constant balancing act of learning and improvising as astronauts adapt to their unique, zero-gravity surroundings. The daily rhythms on the ISS may revolve around scientific research, equipment maintenance, and health checks, but each task takes on a new twist in weightlessness. Even the simplest activities require a shift in thinking and a level of creativity that is unique to life in space.

Navigating the station, for example, becomes an exercise in fluid, three-dimensional movement. Astronauts learn quickly that their initial impulse to walk or push off too hard can send them ricocheting across the modules. Instead, they practice a kind of graceful drifting, using walls, rails, or even an occasional mid-air push-off to glide through the station's corridors. Many remember their first clumsy attempts to navigate, sometimes leading to gentle crashes or unexpected encounters with floating objects. Jack Fischer recalls these early moments with a laugh, admitting to an "unplanned head-on collision" with a tool kit that seemed to have its own trajectory, a humorous reminder of the new physics that govern their every move.

Simple routines, like eating and drinking, become surprising experiences. Without gravity, liquids don't pour or settle, so every meal is eaten with a sense of fascination. Astronauts sip from specially designed pouches to prevent drinks from floating away in droplets, and food is often kept in sealed packets or with a bit of adhesive to keep it attached to a surface. Drew Morgan once shared a story of a tortilla that slipped from his hand mid-meal, lazily drifting out of reach until he could gently chase it down. Even small tasks like brushing teeth or washing up involve creativity. Without running water, they use rinseless wipes and swallow their toothpaste to avoid stray blobs floating through the cabin, adding an unexpected twist to their daily hygiene routine.

Sleeping in zero gravity, too, takes some getting used to. Astronauts are strapped into sleeping bags attached to the walls, with no pillows or bedding, as they float in their sleep quarters. There's a funny but practical reason for the restraint: without it, they'd drift around the cabin like gently tumbling leaves. Karen Nyberg describes the feeling of falling asleep weightless as "oddly comforting," almost like being cradled in mid-air. After the initial disorientation, many grow to love the freedom it offers, enjoying a sleep that feels like floating in a dreamlike state. This adaptation is one of the many ways the body and mind adjust to space, creating a new sense of comfort within an entirely foreign environment.

The adaptability of the human body and mind becomes evident in the way each astronaut adjusts to these routines, transforming unfamiliar experiences into new normals. Over time, even tasks as unconventional as clipping nails in zero gravity—where clippings can float away if not carefully managed—become second nature. This adaptability is also a source of camaraderie and humor, as crew members swap tips on techniques that are quirky but essential for surviving daily life in space.

Each day aboard the ISS becomes a reminder of how the human body and mind can redefine their boundaries in response to any environment, no matter how unconventional. Through patience, a sense of humor, and the ability to find joy in every challenge, astronauts learn to thrive in a world without gravity, turning what was once foreign into something uniquely their own.

Chapter 5: The International Space Station – Humanity's Orbiting Lab

The International Space Station stands as a remarkable testament to human cooperation, an orbiting laboratory born from the shared vision of multiple nations working together to push the boundaries of exploration and knowledge. Its construction required the dedication and resources of agencies from around the world—NASA, Roscosmos, ESA, JAXA, and CSA—all collaborating to bring this ambitious project to life. More than just a scientific achievement, the ISS embodies an ideal of unity, symbolizing what humanity can achieve when it sets aside differences to reach toward a common goal. Floating 250 miles above Earth, it serves as a beacon of possibility, where astronauts of different nationalities live and work side by side, sharing both the challenges and rewards of life in space.

Life on the ISS revolves around scientific discovery, with research covering an extraordinary range of disciplines that benefit life on Earth as well as future space missions. Anne McClain describes the station as a "proving ground" where experiments unfold daily, each contributing to humanity's understanding of biology, physics, chemistry, and beyond. The microgravity environment offers a rare opportunity to study phenomena that behave differently outside Earth's gravity, revealing insights that are otherwise impossible to obtain. In the realm of medicine, for instance, the study of cellular behavior in zero gravity has led to advances in understanding diseases like osteoporosis and muscular atrophy, which are accelerated in space and therefore provide a unique model for researchers.

For astronauts like Nicole Stott, the station is also a powerful observatory, offering an unobstructed view of Earth and its atmosphere, crucial for monitoring changes and collecting data on our planet's health. Research aboard the ISS includes tracking shifts in climate patterns, studying pollution levels, and observing natural phenomena like storms and wildfires from an unparalleled vantage point. These studies provide invaluable data for scientists on the ground, offering a long-term perspective that deepens our understanding of Earth's ecosystems and how human activity impacts them.

The station also hosts studies that prepare for the future of space exploration. Experiments on plant growth in microgravity, for example, are helping scientists develop sustainable food sources for long-term missions, laying the groundwork for journeys to Mars and beyond. The physical and psychological effects of space travel on the human body are closely monitored as well, providing critical information on how extended periods in space affect astronauts' health and well-being. By understanding these effects, scientists are better equipped to create strategies that protect future explorers, ensuring their safety and performance on distant worlds.

To the astronauts who live and work aboard the ISS, this collaborative, international space laboratory is more than just a workplace—it is a symbol of global unity and a glimpse into the potential future of humanity. As Anne McClain puts it, the station represents an unprecedented achievement, where the spirit of exploration transcends borders. The ISS is a bridge to new knowledge and a reminder of the interconnectedness of all life on Earth, uniting people across cultures, languages, and nations in the pursuit of discovery and a better understanding of our place in the cosmos.

For those who spend months aboard the International Space Station, it becomes not only a workplace but a second home, a unique, floating habitat where camaraderie and cooperation define every day. Living in such close quarters with colleagues from around the world, astronauts experience a sense of unity that few other environments can foster. The daily routines, the shared challenges, and the collective achievements create a bond that transcends any one individual's background or nationality. They are no longer representatives of separate countries; they become a single team, united by the shared mission of exploration and discovery.

The ISS itself, an intricate construction pieced together through the combined efforts of multiple nations, embodies this spirit of cooperation. It is a structure built across continents, each module reflecting the expertise, technology, and vision of the agencies that contributed. To the astronauts, the station is a tangible reminder of what humanity can achieve when it works together, a symbol that reinforces the values of peace and collaboration with every orbit around Earth. Anne McClain often reflects on this unity, feeling a deep connection to her crewmates as they work toward common scientific goals and support each other through the everyday challenges of life in space.

As a workplace, the ISS demands constant coordination, where crew members rely on one another to ensure every mission runs smoothly. Each astronaut brings a set of specialized skills, and they train extensively to anticipate each other's needs, trusting in one another's knowledge and abilities. This shared trust strengthens as they adapt to a life where cooperation is not just helpful but essential. Any experiment, repair, or mission task requires seamless communication and a spirit of teamwork that becomes second nature. Nicole Stott recounts moments of spontaneous problem-solving, when a diverse mix of ideas from different cultural and technical backgrounds comes together to overcome challenges that arise miles above Earth.

The environment itself encourages a sense of mutual respect and empathy. Floating side by side in zero gravity, each astronaut gains a deep appreciation for the others' backgrounds, stories, and perspectives. As they look down upon the

Earth, they feel the paradox of their distance from and connection to everything below. They are acutely aware of how small and fragile the planet appears and how vital it is that humanity learns to work together. The world's divisions seem insignificant from their vantage point, giving way to a powerful awareness of unity and a sense of duty to represent that unity on Earth.

In this isolated, orbiting laboratory, the ISS is more than just a hub for scientific progress; it is a bridge linking different parts of the world through shared purpose. Each experiment they conduct, each discovery they make, is a contribution not only to the advancement of science but to the strengthening of global collaboration. In their time aboard the station, astronauts find that it is not only a place of research but a testament to the idea that humanity, despite its divisions, can come together to achieve remarkable things. They carry this sense of unity with them when they return to Earth, hopeful that the cooperation they experienced can inspire others to see beyond borders and work together for the greater good of all.

The ISS, in its role as both a home and workplace, stands as a constant reminder of what can be achieved through collective effort. It is a beacon of scientific progress and human cooperation, a place where astronauts from diverse backgrounds come together to share a common vision—one that reaches beyond our individual worlds and into the vastness of space, reflecting our shared potential and responsibility to each other and to the planet below.

Chapter 6: The Cupola – Earth Observed from Above

The Cupola module on the International Space Station offers a view like no other, a seven-windowed vantage point that reveals Earth in all its stunning detail, an endless tapestry of colors and textures spread out against the darkness of space. For astronauts, the first moment of gazing through the Cupola is unforgettable. They float gently toward the windows, feeling the magnetic pull of the view, a sight that their years of training and imagination could never fully prepare them for. What they encounter is a living planet, swirling with clouds, oceans, mountains, and deserts, all illuminated by the light of the sun. This view, often described as breathtaking, is not just visually beautiful; it strikes deep into the heart, evoking a profound and often emotional response.

For Anne McClain, this first look at Earth was a moment of quiet awe. She recalls being overwhelmed by a sense of scale, where the vastness of the planet and the fine, delicate line of the atmosphere seemed to capture the fragility and interconnectedness of life. Nicole Stott, too, felt an immediate and transformative connection. She would close her eyes for a moment before opening them again, hoping to fix every detail in her memory. As the layers of blue, green, and white shifted below her, she was struck by the sharpness of the view, the vibrant clarity of the colors, and the way the landscape seemed to breathe with each orbit. The experience defied words, a sight so powerful that it moved her beyond intellectual understanding and into a feeling of unity with the Earth itself.

Seeing Earth from this perspective brings a flood of realizations, each one complex and uniquely personal. Many astronauts describe feeling both immense pride and humility. There is a recognition of Earth's beauty but also of its vulnerability, the thin blue line of atmosphere serving as a fragile veil between life and the void of space. Jack Fischer remembers watching storms move across the planet, lightning illuminating the clouds in delicate flashes that looked like pulses of light traveling over the surface. In these moments, Earth seemed alive and dynamic, a single, unified organism with intricate and ever-changing systems.

This view from the Cupola is often a moment of profound introspection, a reminder of everything that makes Earth home. The mountains, forests, and oceans, places the astronauts had visited or hoped to visit, suddenly become pieces of a larger, interconnected whole. There is a sense of both distance and closeness, of being physically removed but emotionally more connected than ever. Astronauts like David Saint-Jacques speak of this view as one that realigns priorities, bringing a new understanding of our shared responsibility to protect the planet. As he looked down at the continents, he could no longer see borders or divisions, only the vast, sweeping continuity of land

and sea, reinforcing the message that humanity shares a single, precious home.

The transformative power of this view leaves a lasting imprint. Many astronauts find that their time spent gazing through the Cupola becomes one of the defining aspects of their journey, a kind of spiritual experience that shapes their perspectives long after they return to Earth. They are left with a renewed sense of purpose, a desire to share this perspective with others, to help people understand the beauty and fragility of our world. In that stillness of space, suspended above the planet, they see Earth not just as a collection of countries or landscapes but as a singular, interconnected entity, a reminder of the unity and resilience that bind all of life. The view from the Cupola becomes, for them, a bridge between the worlds above and below, a reminder of both the awe-inspiring beauty of our planet and the shared responsibility of all who call it home.

From the vantage point of space, Earth reveals itself as a masterpiece of colors, shapes, and textures, a living work of art that surpasses any previous conception of beauty. Through the windows of the space station, astronauts encounter a view that feels almost unreal in its intensity. Earth, bathed in sunlight, appears as a swirl of vibrant blues, deep greens, earthy browns, and snow-capped whites, each color more vivid and striking than imagined. The landscapes, free from borders or human divisions, blend seamlessly into one another, creating a mosaic that stretches out infinitely beneath them.

Jessica Meir, seeing Earth from this view for the first time, describes the experience as overwhelming, a "sensory overload" that filled her with both admiration and humility. The colors were brighter and more dynamic than she'd ever seen, with the oceans presenting a gradient of blues that transitioned from turquoise in the shallow reefs to nearly black in the deepest waters. The sun's reflection on these surfaces created glimmers of light, dancing on the waves like jewels against the backdrop of vast, open seas. Meir recalls being captivated by the sunlit edges of continents and islands, their borders softened by clouds, each scene changing and evolving with every orbit.

For David Saint-Jacques, the view evoked a sense of life, movement, and unity. He was fascinated by the sight of mountain ranges, sharp and ancient, cutting across continents like natural sculptures. He could see the grand deserts, their sands forming delicate patterns that looked as though they had been painted by a careful hand. There were forests stretching endlessly, rivers snaking through valleys, and cities that appeared as faint clusters of light. From space, he observed the texture of these landscapes, each one unique, each telling its own story. In orbit, the Earth's curvature became a constant reminder of its shape, its finiteness, and the thin, delicate line of atmosphere surrounding it—a veil that felt both fragile and enduring.

The colors of Earth at sunrise or sunset are especially captivating, a display that defies the limits of photography. Jessica Meir remembers watching as the sun dipped below the horizon, casting the atmosphere in layers of orange, pink, and purple, with shadows stretching over the planet's surface. This daily event became a treasured ritual for many astronauts, a moment of profound peace and reflection, where they felt the rhythm of Earth even from miles above. The experience of watching a sunrise from space, where the sun rises or sets every 90 minutes, is both grounding and surreal, reminding them of the passage of time and the resilience of life below.

Seeing Earth from this distance is transformative, inspiring astronauts with a new understanding of our planet's uniqueness and interconnectedness. They speak of the way the textures and colors shift with weather patterns, how thunderstorms explode in white flashes across the surface, and how the gentle curves of coastlines blend into the surrounding oceans. This view gives them a perspective that is nearly impossible to attain on the ground, a perspective that reveals the Earth not just as a place, but as a delicate and intricate system of life. It is a shift that goes beyond words, affecting their perception of humanity and their role in preserving this extraordinary home.

This new outlook becomes an essential part of who they are. Many astronauts, upon returning to Earth, describe feeling more connected to the planet and its people, more aware of its beauty and its fragility. The view from space shows them a world without boundaries, a seamless blend of ecosystems and landscapes that sustain life. In these moments of reflection, they recognize that they are part of something larger, a collective responsibility to protect and cherish what they've seen. The experience of seeing Earth from above does more than inspire awe; it instills a deep sense of purpose and a desire to share this perspective, hoping to bring a piece of that transformative view to those of us on the ground.

Chapter 7: A World Without Borders – The Unity of Humanity

From the heights of orbit, Earth appears as a continuous, interconnected whole, a living mosaic undivided by the lines that humans have drawn. Looking down from the vastness of space, astronauts are struck by the absence of borders, the political boundaries and distinctions that otherwise define maps and globes. Instead, they see land and sea, desert and forest, mountain and valley—all seamlessly flowing into one another, uninterrupted and unified. The continents blend gracefully, each ecosystem merging naturally into the next, forming a complete picture of the planet as it truly is: whole, cohesive, and resilient.

Tracy Dyson recalls this perspective as both liberating and humbling, her view of the Earth stripped of the artificial separations we use to define ourselves. She describes how, from space, she could gaze across countries and continents, seeing them not as territories but as interconnected pieces of a global landscape, each supporting the other in a delicate balance. This view challenged her perception of division, reminding her that the things that bind us far exceed what separates us. Earth, from her vantage point, was not a collection of nations but a single, boundless home, a shared environment that supports all life.

Don Pettit also reflects on this experience, recounting how, as a child, he had memorized maps with countries distinguished by bright colors. Yet, as he looked down from space, he was struck by how different the reality was. There were no lines visible from his perch above the atmosphere, no colors marking nations—only the natural curves of coastlines, rivers winding through valleys, and forests sprawling across continents. The distinctions that once seemed so clear now felt inconsequential, almost arbitrary. He felt a sense of unity with all life on the planet, an understanding that humanity exists as part of an interdependent whole, where each region and each ecosystem contributes to the survival and health of the rest.

This perspective brings a realization of both the beauty and the vulnerability of Earth. Without borders to divide it, the planet's interconnectedness becomes undeniable. Astronauts see the continuous flow of weather systems, the ripple of storms across the ocean, the way pollution drifts across regions, affecting air and water far beyond its origin. Tracy Dyson notes how this view reshapes priorities, shifting from national interests to a shared responsibility for protecting the planet. The realization is powerful: every environmental impact, every decision made, resonates across invisible lines, affecting people and ecosystems on the other side of the world.

Observing Earth from space allows astronauts to see humanity in its entirety, to view us all as members of a single, fragile world. This view does not diminish their love for their own countries; rather, it expands their understanding, reminding them that all humans share the same land, water, and sky. The awareness that they gain—that Earth is a singular entity with a shared future—leaves a lasting impression. It instills a sense of unity and compassion that transcends borders, fostering a vision of humanity that emphasizes cooperation, empathy, and a collective duty to care for the planet.

The experience of seeing Earth without boundaries offers a profound shift in perspective, revealing the insignificance of our divisions when compared to the importance of our shared home. It is a perspective that each astronaut carries with them upon their return, a constant reminder of Earth's beauty, vulnerability, and the interconnectedness that binds us all. In their reflections, they often hope that more people could see the world as they have, believing that this view might inspire a deeper respect for the planet and a renewed commitment to safeguard it—not for any one nation or people, but for all.

As astronauts gaze down at Earth from space, they come to a profound realization: humanity, with all its diversity and complexity, is fundamentally one. The differences that so often divide us—whether cultural, political, or ideological—seem to fade away against the backdrop of Earth's breathtaking beauty and unity. Suspended above the planet, they see only a single, interconnected world, vibrant and alive, where oceans, forests, mountains, and cities all weave together into a seamless whole. This perspective strips away the everyday distinctions that typically define human experience, leaving a deep sense of interconnectedness and shared existence.

For many astronauts, this view is transformative. Jack Fischer describes the moment as one of overwhelming unity, a feeling of belonging to a collective humanity that transcends borders. Floating miles above the surface, he is filled with a sense of awe for the intricate web of life below, an ecosystem where every person, every plant, every animal is part of a larger tapestry. In the stillness of space, he sees not the labels or boundaries that divide, but the invisible threads that bind all of life, a reminder that, at our core, we are all inhabitants of the same planet.

This realization of unity often reshapes how astronauts perceive their purpose. They no longer see themselves as representatives of one nation or one mission but as part of a greater human endeavor. Nicole Stott recalls moments of quiet reflection as she watched Earth from orbit, feeling connected to everyone she knew—and even to those she didn't. The view offers a perspective that makes the challenges, achievements, and struggles of humanity feel collectively shared, inspiring a compassion and empathy that is difficult to describe yet unmistakably real. She observes that from this vantage point, the boundaries that separate people dissolve, replaced by a profound respect for the diversity that makes humanity rich and resilient.

The interconnectedness of Earth is visible not only in its landscapes but in the rhythm of life below. Watching lightning storms sweep across continents, seeing weather patterns form and dissipate, and observing natural phenomena like hurricanes and wildfires highlight the unity of Earth's systems. What begins in one place impacts another; oceans influence climates, forests regulate air, and winds carry seeds, rain, and even pollution across continents. This natural interconnectedness reinforces the idea that every part of the planet plays a role in sustaining life, a delicate balance that requires harmony and cooperation to thrive.

For astronauts, this perspective is both humbling and empowering. In orbit, they are reminded of humanity's shared fate, of the responsibility that each of us holds to protect and preserve the planet. Don Pettit reflects on how space travel illuminates the simple truth that, regardless of our backgrounds, all humans breathe the same air, rely on the same resources, and share a common desire for survival and well-being. From space, he sees Earth as a single, fragile entity, floating in the vastness, our only home, and the ultimate unifier of human experience.

This view inspires a vision of humanity united in purpose, where our differences enrich us rather than divide us. It serves as a reminder that each of us plays a role in the larger story of Earth, that every action ripples outward, impacting not just our immediate surroundings but the world as a whole. The perspective offered by space travel is one that reveals the strength and beauty of unity, a reminder that our shared existence on this planet is far more significant than the divisions that often separate us.

In their reflections, many astronauts express a hope that others could share in this view, believing that it would inspire a broader understanding of our place within the universe and a deeper commitment to protect the Earth. This unifying vision becomes a permanent part of their outlook, a reminder of the interconnectedness that defines life on Earth and the humanity that binds us all.

Chapter 8: Earth's Dynamic Beauty – The Colors and Patterns of Nature

From the vantage point of space, Earth's landscapes are a living canvas, constantly shifting in color, texture, and light. The view from the space station reveals a planet in motion, its surfaces and skies alive with patterns that are mesmerizing in their complexity. As astronauts watch, they witness vibrant shades of blue and green from the oceans, vast sweeps of amber and ochre from deserts, and rich forest greens that hint at the density of life below. The hues of Earth are intense, each landscape marked by its distinct palette that changes with seasons, weather, and the angle of the sun. Seen from above, these colors appear vivid and pure, a reminder of nature's ability to paint the planet with strokes of unparalleled beauty.

The dance of light and shadow across Earth's surface adds a dramatic depth to the view, highlighting mountain ranges, valleys, and coastlines in ways that ground-based perspectives could never capture. As the sunlight shifts, casting shadows across plains and ridges, the landscapes take on new dimensions, giving astronauts an almost three-dimensional perspective of the planet's terrain. Shadows stretch long across the deserts, and the curve of the atmosphere bends the light into delicate gradations of color during sunrise and sunset, wrapping Earth in halos of pink, purple, and gold. For astronauts, these moments are like a continuous work of art in motion, a masterpiece created by nature's hand.

Some of the most awe-inspiring sights are the immense, swirling storms that move across oceans and continents. From space, hurricanes appear as giant spirals, tightly coiled with layers upon layers of clouds that converge at the eye of the storm. Jack Fischer recalls his amazement at seeing the formation of a hurricane, its sheer size and power humbling even from miles above. The eye of the storm, dark and foreboding, lies at the center of a churning vortex of white clouds, rotating in a slow yet relentless spiral. In these moments, astronauts are reminded of the raw force of nature, a reminder of Earth's power and the delicate balance that sustains life.

The view also reveals fleeting, powerful displays of natural phenomena like lightning. At night, thunderstorms illuminate the planet in electric flashes, bursts of light that travel across the clouds like synapses firing in a brain. Jessica Meir describes watching lightning storms over vast areas, the clouds below lighting up in staccato bursts, each strike a reminder of the energy coursing through Earth's atmosphere. These storms can stretch over entire continents, their flashes visible for miles, illuminating everything from dense forests to open plains. From above, lightning appears almost as an abstract pattern, tracing connections across the skies and lending the planet an otherworldly beauty.

Sandstorms, too, are mesmerizing to observe from orbit. When vast clouds of dust are lifted from the deserts of the Sahara, they form enormous, sweeping plumes that stretch across the ocean, sometimes reaching as far as the Americas. Reid Wiseman recalls seeing one of these sandstorms for the first time, astonished at the sheer scale of the dust cloud as it moved across continents, transforming the blue skies into shades of brown and tan. This dust, rich with minerals, plays a role in ecosystems far from its origin, nourishing soils and impacting climate in distant regions, a reminder of Earth's interconnected systems at work.

Each of these phenomena—the storms, the sand, the light, and the shadows—reveals the dynamic, ever-changing nature of Earth, where landscapes and weather patterns shift in a continuous cycle of transformation. For astronauts, watching these events unfold provides a powerful connection to the planet's rhythm, a realization that Earth is not static but alive with energy, movement, and life. Every view, every turn of the planet, offers something new, something extraordinary, as the ever-changing landscapes of Earth remind them of the resilience and beauty of the world they call home.

From the orbiting heights of space, astronauts experience Earth as a vibrant, dynamic canvas, alive with colors, textures, and patterns that shift with each passing hour. Gazing down, they see the planet not as a static sphere but as a masterpiece in constant transformation, its beauty revealed through intricate details that come alive under varying light and shadow. The landscapes below—mountains, rivers, deserts, and coastlines—take on sculpted forms, with sunlight accentuating their textures and creating a visual depth that is breathtaking and humbling. The surface of Earth tells a story, one painted by natural forces over millennia and continually refreshed with each dawn and dusk.

One of the most striking features of Earth from space is the way light interacts with the atmosphere and landscape. As the sun rises and sets, casting its golden glow across the surface, astronauts witness a layered play of colors, from deep reds and purples to bright yellows and oranges, that shift as the angle of the light changes. These moments reveal details of the planet's surface—the sharp outlines of mountain ranges, the shadowed valleys, and the contours of ridges and plains. The interplay of light and shadow brings a richness to the view, transforming landscapes into scenes of incredible depth and complexity. For astronauts, these moments create a sense of intimacy with Earth, as if they are witnessing the planet's quiet, natural rhythms from afar.

Weather patterns further enhance Earth's artistry, showcasing the planet's energy and the complex systems that sustain life. Storms, with their swirling clouds, create visual spectacles as they move across continents, bringing rain, wind, and change. Hurricanes, viewed from above, appear as perfect spirals of cloud bands converging on a central eye—a mesmerizing formation that reveals the raw power contained within the atmosphere. These immense storms unfold in slow motion, and their reach, stretching over hundreds of miles, is both beautiful and intimidating. For astronauts, watching a hurricane's structure from above is a reminder of nature's immense force and a testament to the planet's inherent power.

Lightning storms are another example of Earth's dynamic energy on display. At night, these storms reveal themselves through sudden, illuminating flashes, casting brief, ghostly glows across the clouds. Seen from space, lightning resembles an abstract dance of light, each bolt arcing across the cloud tops, a split-second expression of energy moving across the sky. The spectacle often spans vast distances, with clusters of lightning strikes moving in sync as the storm travels. For astronauts, this display is one of nature's most dramatic phenomena, a reminder of the electricity and energy constantly pulsing through Earth's atmosphere.

Sandstorms, too, present a unique visual from space, appearing as vast plumes of dust lifted from deserts and carried over oceans. From the Sahara, massive dust clouds emerge, their tan and golden hues sweeping across the Atlantic, occasionally reaching the Americas. The sight of these storms stretching across continents illustrates Earth's interconnected systems, where particles from one region nourish soils in another, bridging continents in ways often invisible from the ground. This transport of dust is both a reminder of Earth's resilience and a testament to the delicate balance that sustains its ecosystems.

These scenes—from sculpted mountain ranges to swirling storms and drifting sand—showcase Earth's artistry, revealing a planet that is not only powerful but also vulnerable. Astronauts often describe a sense of reverence as they observe these natural displays, recognizing the harmony and balance required to sustain such a world. They see firsthand the thin, fragile line of atmosphere that wraps the planet, the only shield between life and the harshness of space, underscoring Earth's delicate place in the universe.

For those who have seen Earth from space, this perspective instills a deep appreciation for the beauty and complexity of the planet, along with a sense of responsibility to protect it. The artistry of Earth is more than visual beauty; it is a story of interconnected systems, each part essential to the whole. As astronauts return to Earth, this vision stays with them—a reminder of the planet's power, its beauty, and its profound vulnerability. They bring back with them an understanding that the Earth, like any masterpiece, must be cherished and safeguarded for future generations to witness and appreciate.

Chapter 9: Practical Use Cases and Setup Tips

From the heights of space, astronauts observe some of Earth's most magnificent natural forces on display, each a powerful reminder of the planet's vibrancy and vitality. The auroras—those ethereal, shimmering curtains of light—are among the most captivating sights. As solar particles collide with Earth's atmosphere, they create luminous waves of green, blue, and red that dance along the polar regions. Seen from orbit, the auroras flow like rivers of light, reaching up toward space in a graceful, pulsating rhythm. These displays remind astronauts of the immense energy flowing through the Earth's magnetic field and its role in shielding the planet from the sun's radiation.

Oceans, covering most of the Earth's surface, reveal a quieter, yet equally powerful force. Astronauts describe how vast stretches of water take on different hues under varying light, shifting from deep, almost-black blues to turquoise and jade near coral reefs. Waves and ocean currents, normally invisible to the eye, become evident as patterns and textures that ripple across the water's surface. Storms form over these oceans, gathering moisture and strength, and eventually grow into massive systems that make their way toward land. From above, the interconnectedness of the oceans is unmistakable, a continuous force that shapes weather patterns and stabilizes climates.

Thunderstorms are another dramatic reminder of Earth's natural energy. At night, these storms look like flickering constellations scattered across continents, with lightning illuminating clouds in a ghostly glow. Viewed from space, lightning storms are beautiful but haunting—a dance of electricity that lights up the darkness. The sheer scale of a storm system is humbling, as astronauts watch lightning arcs trace paths over forests, cities, and plains below. Jack Fischer recalls how mesmerizing it was to watch lightning flashes within vast storm clouds, almost like watching neurons firing in the brain. These storms serve as powerful reminders of the Earth's dynamic atmosphere and the volatile systems at work within it.

Despite the majesty and intensity of these forces, astronauts are keenly aware of Earth's fragility. The atmosphere, which enables life and protects the planet from the void of space, appears from orbit as a thin, delicate veil, a sliver of blue wrapped around the globe. This fragile line is all that stands between Earth and the harsh radiation and vacuum beyond. For those observing from space, the atmosphere's thinness is startling; it feels vulnerable, almost as though a single disruption could unravel it. They realize that this atmospheric layer, barely visible from their vantage, is critical to sustaining life and that its composition, temperature, and resilience are all finely balanced.

The auroras, thunderstorms, and oceans highlight the raw, unchecked power of nature, yet these forces all depend on the stability of Earth's atmosphere. Observing the planet from space, astronauts often reflect on how dependent life is on this thin layer, which nurtures ecosystems, regulates temperatures, and shields Earth from cosmic hazards. Nicole Stott describes this awareness as a "wake-up call," a realization that the atmosphere's fragility makes it essential to care for the systems that maintain its integrity.

This vantage point, revealing both the grandeur of Earth's natural forces and the fragility of its atmosphere, leaves astronauts with a lasting sense of responsibility. They return with an understanding that life on Earth is protected by a system of interconnected elements, each one vital to the whole. As they gaze down at the storms, oceans, and auroras, they are filled with both awe and respect, recognizing that while nature's power is immense, the protection of life is ultimately fragile. This perspective fuels a desire to protect and preserve Earth's atmosphere and the delicate balance that sustains all life, so that future generations may continue to witness and marvel at the beauty of a planet alive with power and resilience.

From the vantage point of space, Earth's beauty is profound, but its resilience is matched by a striking vulnerability that is impossible to ignore. Watching the planet from above, astronauts gain an intimate view of the systems that make life possible, from the swirling clouds and dynamic oceans to the thin, protective layer of atmosphere that cocoons Earth against the harshness of space. This perspective, so unique and distant, reveals the planet as both enduring and fragile, a living entity where each part supports the whole in a delicate balance. The spectacle of Earth's natural displays—the play of auroras at the poles, the vast stretch of oceans, and the power of storms—reminds them that our planet, while resilient, requires protection and care.

Auroras, with their shimmering lights reaching toward the stars, showcase the interaction between Earth's magnetic field and the solar winds, a dynamic that shields the planet from harmful cosmic radiation. Watching this cosmic dance from above, astronauts recognize the role that Earth's magnetic field plays in maintaining this shield, which, like the atmosphere, is a critical defense against external forces. Observing this phenomenon instills a heightened awareness of the planet's natural protective systems and the intricate balance that sustains life. The sight of these auroras, so beautiful and yet so functional, underscores the importance of preserving the planet's natural shields—its atmosphere and magnetosphere—without which life would be vulnerable to threats from space.

The vast oceans below, absorbing sunlight and powering weather systems, also highlight Earth's remarkable yet delicate equilibrium. Astronauts observe how ocean currents regulate temperatures and support diverse ecosystems, carrying warmth from the equator toward the poles and impacting climate patterns across the globe. From space, they see how changes in one part of the ocean can ripple outward, affecting distant regions in ways that might go unnoticed from the ground. The interconnectedness of these bodies of water emphasizes Earth's reliance on stable climates and clean oceans. Witnessing this unity from space fosters a sense of duty among astronauts to protect the oceans from pollution, warming, and acidification—threats that disrupt these natural processes and endanger the delicate marine environments that sustain so much of life on Earth.

Thunderstorms, hurricanes, and other powerful weather patterns are mesmerizing in their strength, but they also reveal the planet's sensitivity to change. These systems, fueled by temperature differences in the oceans and atmosphere, are visible reminders of Earth's energy in motion. Yet the changing frequency and intensity of such storms hint at larger shifts in climate, reminding astronauts of the consequences of human impact on the environment. Observing the scale and frequency of these storms from space drives home the reality of climate change, as the planet responds to shifts in temperature, pollution, and land use. The power of these storms, juxtaposed against the thinness of Earth's atmosphere, makes the need for sustainable practices on the ground ever more pressing.

For many astronauts, seeing Earth as a unified system sparks a profound environmental awareness. They return to Earth with a renewed commitment to advocacy, inspired by the sight of the planet as a small, finite sphere in the vastness of space, vulnerable to the effects of human activity. This view instills a sense of duty, as astronauts realize that Earth's resilience, though remarkable, is not boundless. It depends on humanity's respect for the natural systems that keep it in balance. The thin blue line of atmosphere that encircles the planet, visible from space, becomes a symbol of this fragility, a reminder of how dependent life is on maintaining this delicate layer.

The view from above fosters a global perspective, making borders, divisions, and individual concerns seem small in comparison to the overarching need to protect Earth as a shared home. Astronauts often describe a sense of stewardship, an understanding that caring for the planet is not just a personal or national responsibility but a duty to humanity as a whole. The resilience of Earth, tempered by its vulnerability, serves as a powerful reminder that every choice we make impacts the balance that sustains life. This perspective, gained from the solitude and distance of space, brings them back to Earth as advocates, inspired to protect the beauty, power, and delicate equilibrium of the planet they were privileged to see from afar.

Chapter 10: Troubleshooting and FAQs

Leaving the International Space Station is an experience layered with emotion, a bittersweet parting from a place that has become both a workplace and a home. Astronauts often speak of their final days aboard the station with a sense of reflection and gratitude, aware that they are about to leave behind a vantage point that very few have known. The ISS has been their refuge, a place where they witnessed Earth from above, experienced weightlessness, and forged deep connections with their fellow crew members. As they prepare for the journey back, there is an undeniable pull toward Earth—a longing for family, familiarity, and the sensory experiences of home. Yet, there is also a profound sadness at leaving the quiet vastness of space and the routine that has become a cherished part of their daily lives.

The descent begins with the intense re-entry sequence, a dramatic reintroduction to gravity as their capsule plunges through the atmosphere.

After months of floating freely, feeling weight return to their bodies is almost jarring, a visceral reminder of the planet's pull. The journey is marked by powerful physical sensations as the G-forces press them back into their seats, bodies once again anchored by gravity's grip. Many describe this phase as both exhilarating and humbling—a welcome return to Earth and a reminder of its grounding force, yet a goodbye to the freedom of space.

Upon landing, re-adapting to Earth's gravity is a challenge that requires both physical strength and mental resilience. The body, having spent extended time in a zero-gravity environment, is initially unprepared for the demands of weight. Muscles that once moved effortlessly in weightlessness now ache under even the simplest movements. Walking, lifting, and balancing are no longer automatic actions but deliberate efforts. Some astronauts find themselves momentarily unsteady, their sense of equilibrium readjusting to the ground beneath their feet. Bones and muscles, having deconditioned in space, require time to regain their strength. It's a humbling experience, as they relearn to navigate their own bodies in a familiar, yet suddenly challenging, environment.

Psychologically, the transition back to Earth can be equally complex. Space has a way of expanding the mind, of offering a perspective that fundamentally changes one's sense of place and purpose. Many astronauts experience a shift in their personal identity, carrying with them the awareness of Earth's interconnectedness and fragility. The daily concerns and routines of life on Earth can seem trivial in contrast to the vastness of space and the profound reflections it inspired. Astronauts often struggle to articulate these feelings to those who have not shared the experience, grappling with a sense of solitude in the very world they've returned to.

The return to Earth also stirs a heightened appreciation for the simple pleasures of home—the warmth of sunlight, the feel of grass underfoot, the colors of a sunset, and the sound of a loved one's voice. Every sensory experience, every natural element, feels amplified, grounding them in a newfound gratitude for Earth's beauty and vitality. This re-sensitization to everyday life serves as a reminder of the preciousness of home, a perspective deepened by the privilege of having seen it from afar.

In the months that follow, many astronauts reflect on the transformative impact of space on their personal identity. They carry a new sense of purpose, often feeling compelled to advocate for environmental protection, scientific exploration, and international collaboration. The memories of Earth from above, seen without borders, become a lens through which they view their role in humanity's future, driving them to contribute positively to the world they returned to. This experience becomes an inseparable part of who they are, a call to share the insights they gained and to inspire others to care for our shared home.

For those who have ventured into space, returning to Earth is not simply a physical re-entry but a profound re-connection with life below. They come back with a perspective shaped by the stars, touched by the vastness of the cosmos, and infused with a sense of unity that remains with them, defining their identity and purpose long after they have left the International Space Station.

Upon returning to Earth, astronauts find themselves re-immersed in a world that, although familiar, now feels profoundly different. The simplest aspects of daily life take on a vivid significance, as if they're experiencing them for the first time. After months in space, where colors are muted, textures are artificial, and the landscape is limited to metal, glass, and the occasional view of Earth below, the sensory richness of life on the planet becomes overwhelming and beautiful. Fresh air, the warmth of sunlight on their skin, the smell of rain, and the vastness of open skies—all of it feels like a rediscovered treasure.

Astronauts often speak of their first steps outside after returning, how the sensation of grass underfoot or a breeze brushing past feels almost surreal. Many find themselves marveling at the diversity of colors in nature, from the deep greens of forests to the fiery hues of a sunset. Every scene is more vivid, more meaningful, as if Earth itself has been painted with greater detail. For some, the sound of birds or the rustling of leaves becomes music, a natural symphony they had missed in the sterile silence of space. These small, everyday encounters—previously taken for granted—now serve as powerful reminders of Earth's beauty and complexity, grounding them in a deep sense of connection to their home.

This renewed appreciation extends beyond the aesthetic, fostering a profound sense of stewardship for the planet. Having seen Earth from space—a small, fragile sphere floating in the vastness—they understand more deeply how interconnected and delicate it truly is. The view of Earth's thin atmosphere, barely perceptible from their orbital perch, is a constant reminder of the planet's vulnerability. This experience changes them; it instills a responsibility to protect the systems that sustain life. The air we breathe, the water we drink, the ecosystems that support countless species—all of these now seem even more worthy of preservation.

Many astronauts return with a heightened awareness of environmental issues, feeling compelled to advocate for sustainable practices and global cooperation. The boundaryless view from space highlights how pollution, deforestation, and climate change do not respect borders; they affect the planet as a whole. With this understanding, they often find themselves engaged in efforts to protect the environment, to communicate the urgency of climate action, and to inspire others to see Earth as a shared home that requires care from everyone.

This sense of responsibility also extends to humanity itself. Having viewed Earth without the divisions of countries or regions, astronauts come back with a sense of unity, a belief that all people are connected by the shared experience of life on this small blue planet. The conflicts and struggles that once seemed large and insurmountable now feel smaller in the context of the world they've seen from space. Many find themselves championing peace, compassion, and collaboration, recognizing that Earth's future depends on humanity's ability to work together, respect each other, and prioritize the well-being of the planet over individual interests.

The lasting impact of space travel on their lives goes beyond mere memories; it reshapes their outlook on existence. The view from space becomes a lens through which they now see the world, a perspective that colors every decision, every action. They carry this newfound reverence for Earth with them, using their experiences to advocate for a better, more sustainable future. The journey into space gives them a gift of understanding, a reminder of Earth's extraordinary resilience and fragility, and a call to protect it for future generations.

In every leaf, every ocean wave, every mountain and forest, astronauts see a reflection of the planet they viewed from above, a reminder of its power and beauty—and its need for care. This renewed appreciation transforms into a lifelong commitment, a dedication to fostering awareness and encouraging others to value and protect our home. For those who have seen Earth from space, the experience becomes an enduring part of their identity, a call to stewardship that continues to shape their lives and inspire those around them.

www.ingramcontent.com/pod-product-compliance
Lightning Source LLC
Chambersburg PA
CBHW070116230526
45472CB00004B/1281